Fossils: Records of History

Harcourt
SCHOOL PUBLISHERS

Orlando Austin New York San Diego Toronto London

Visit *The Learning Site!*
www.harcourtschool.com

A History of Earth

Earth has not always looked as it does now. Scientists believe that Earth and the organisms that live on it have changed over time.

Some scientists theorize that Earth formed from a cloud of gas and dust about 4.5 billion years ago. They also theorize that about a billion years after Earth formed, the first living things appeared. These were single cells. Then over millions of years other living things developed.

Scientists use clues from fossils to learn about changes on Earth that happened long ago. **Fossils** are the preserved remains or traces of past life found in some rocks. Looking at fossils helps scientists tell how life on Earth has changed.

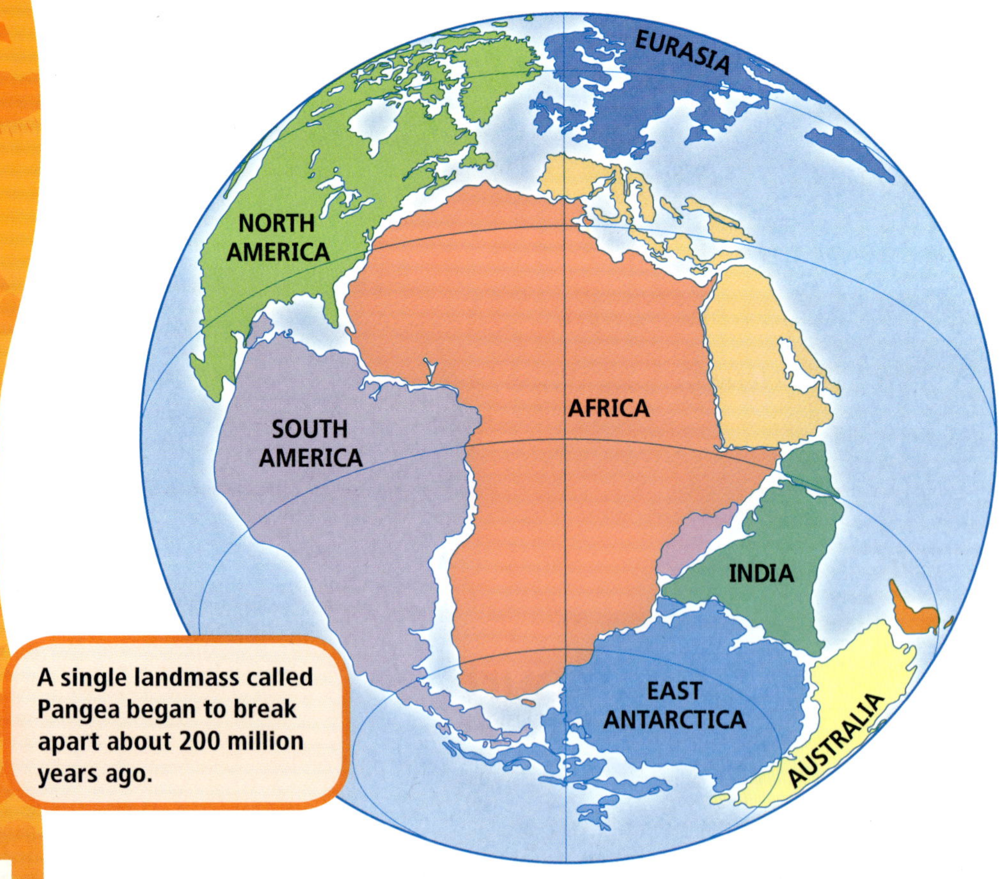

A single landmass called Pangea began to break apart about 200 million years ago.

Studying fossils provides clues to changes to Earth's surface. Fossils that are similar to one another have been found on both sides of the Atlantic Ocean. Some of them come from organisms that could not have crossed the ocean. Finding similar fossils in places far from each other helps support the idea of continental drift. Continental drift is the theory of how Earth's continents move over its surface. Scientists believe that about 225 million years ago all land on Earth was joined together. Scientists call this large landmass Pangea. Evidence suggests that Pangea broke into two pieces, and then into smaller ones, forming the continents known today.

SEQUENCE **How has Earth changed from 4.5 billion years ago?**

Similar fossils, such as this coral fossil, have been found in Africa and South America, providing evidence supporting continental drift.

How Some Fossils Form

What happens when a plant or animal dies? The soft parts quickly decay or are eaten. Hard parts, such as bones, teeth, and shells, last longer. These parts are more likely to remain and become fossils.

Many fossils are found in sedimentary rock. These fossils form when an organism's remains become buried in sediment. The remains may form into fossils if the sediment is not disturbed as it becomes rock. Sometimes fossils in sediment are formed from imprints such as an animal's tracks or the outline of a leaf. These fossils are called trace fossils. Trace fossils can be used to estimate the size of an organism.

The fossil pictured here is the footprint of a dinosaur. Scientists have learned a lot about dinosaurs from studying fossils.

The mold of the shell is the hollow space in the rock.

Molds and casts are two other types of fossils that may be found in sedimentary rock. Both molds and casts form from the hard parts of living things. Suppose the hard part of a living thing gets buried in sediment, and the sediment turns to rock. Sometimes the hard remains dissolve, and a hollow space is left. This is a **mold**. Other times, the hard remains dissolve, a mold forms, and then the mold is filled in with minerals that harden. This is a **cast**.

SEQUENCE **What are the steps that occur when a cast forms?**

Other Fossils

Sometimes entire organisms or their parts are preserved. One way this can happen is when remains become petrified. Petrified means "turned to stone." How do living things become petrified? Minerals may slowly seep into the remains of living things, such as the bones of animals or the woody parts of plants. The minerals harden as they replace the remains of the organism. A rocklike fossil is formed.

Have you ever seen petrified wood? It forms when minerals seep into the wood, taking the place of once-living parts.

An entire animal may also be preserved by becoming trapped in tree sap that hardens. Hardened tree sap is called amber. The remains of insects, spiders, and lizards have been fossilized in this way.

Sometimes entire animals have been preserved by being frozen. The fossilized remains of animals called mammoths have been found in ice and frozen ground in northern parts of Asia and North America. Bones, hair, skin, flesh, and even internal organs have been preserved.

The Mammoth Site Museum in South Dakota is one of the world's largest mammoth exhibits and research centers. Since 1974, fossils of 53 mammoths have been identified.

You probably know that mammoths are now extinct. Many animals and plants that lived in the distant past no longer exist. Dinosaurs are another type of animal that is now extinct. Many other organisms, such as types of ferns, trees, insects, and birds, do not exist and will never live again. Studying fossils helps scientists learn what these organisms were like.

COMPARE AND CONTRAST How are fossils of animals frozen in ice different from other types of fossils?

Fast Fact

Did you know that the mammoth is a relative of modern elephants?

The Age of Rocks

Have you ever seen the Grand Canyon? When you are floating down the Colorado River in the deepest part of the canyon, you are looking at layers of rock that are nearly 2 billion years old. The Grand Canyon was formed as the Colorado River slowly eroded rock layers over time. The river managed to cut through about 20 layers of rock. Looking at the layers of rock is like looking into a slice of Earth's history.

The oldest rocks are at the bottom of the canyon. The youngest rocks are at the top of its walls.

Rock layers can be changed or disturbed by many Earth processes.

Remember that sedimentary rock forms in layers. The oldest layers form first. Each layer that follows is younger than the layer below it. From the position of rock layers, scientists can infer the relative ages of the rocks. The relative age of a rock is that rock's age compared with the ages of other rocks.

Using the position of rock layers to determine the relative age of rocks is not always helpful. You probably know that the positions or appearances of rock layers can change. For example, earthquakes can push rock layers into different positions. Melted rock from deep within Earth may push up into cracks in layers of rock, changing their appearance. Also, wind and water work to erode some layers of rock and deposit new ones.

Focus Skill **COMPARE AND CONTRAST** **How are the rocks near the bottom of the Grand Canyon different from the rocks at the top?**

The Rock Record

Scientists can use index fossils to help them determine the age of rocks, when the position of rock layers has been disturbed. An index fossil can be used to find the age of a layer of rock that is out of order. **Index fossils** are the remains of organisms that lived for a short time in many places on Earth. Index fossils allow scientists to match rocks of the same age found in different places, because they formed at one particular time.

Index fossils help determine the absolute age of rocks. Absolute age tells the age of a rock in years, unlike relative dating.

Scientists have studied rocks to determine the age of Earth. The oldest rocks on Earth are about 4.3 billion years old. Scientists think Earth's oldest rocks have been destroyed over time, so they estimate the age of Earth to be about 4.5 billion years.

To find a rock's absolute age, scientists examine the chemicals found in different rocks and fossils. The numbers in this picture show the years the rocks and soil were deposited.

Fast Fact

There are elements in rocks. Some elements change over time. Knowing how long it takes these elements to change helps scientists find the absolute age of a rock or fossil.

Fossils of organisms that lived in water help scientists determine where lakes, rivers, and oceans once existed.

Fossils can tell about past environments on Earth. The remains of ocean animals have been found on land. This suggests that seas once covered where they lived. Evidence from fossils has also been used to learn that parts of the world have gone through climate changes. Fossils of ferns and reptiles have been found near Earth's poles. This demonstrates that the areas were once warmer than they are now.

CAUSE AND EFFECT **Why are index fossils useful in determining the absolute age of rocks?**

Animals: Then and Now

Fossils show what animals were like in the past and how they have changed through time. Scientists who study fossils and compare them to living things today are called *paleontologists.* Some animals have not changed that much over the years. Other animals are now extinct.

Have you ever seen a horseshoe crab? Looking at fossils of horseshoe crabs, scientists have learned that this animal hasn't changed much through time. Scientists have found fossils of ancient horseshoe crabs that are more than 350 million years old. The picture shows a fossil of a horseshoe crab and a modern horseshoe crab on a beach.

The horseshoe crab is not really a crab. Very similar organisms have been found as fossils.

Studying fossils sometimes shows that the ancestors of modern animals were quite different. For example, the fossils of horselike animals show that ancient horses were different from modern horses. Ancient horses were smaller.

Saber-toothed cats were carnivores with large, sharp canine teeth. They went extinct about ten thousand years ago. Scientists think that saber-toothed cats were about the size of modern lions, but twice as heavy. Some types of saber-toothed cats were ancestors of the modern house cat.

COMPARE AND CONTRAST **How are some modern animals different from their ancestors?**

Hundreds of thousands of saber-toothed cat fossils have been found in the La Brea tar pits of California.

Plants: Then and Now

There is less fossil evidence of plants than of animals. Because plants do not have hard parts such as shells or bones, they often decay too quickly to become fossils.

Most plant fossils that exist are carbon film fossils. A carbon film is a thin coating of carbon that is left behind by a living thing. All organisms contain carbon. Sometimes when an organism is buried, heat and pressure force the remains out, leaving only a thin film of carbon. A carbon film fossil can show a living thing (or some part of it) in detail.

Carbon films of plants show details of roots, stems, and leaves. They also show that plants, like animals, have changed over time.

COMPARE AND CONTRAST **Why are there fewer fossils of plants than animals?**

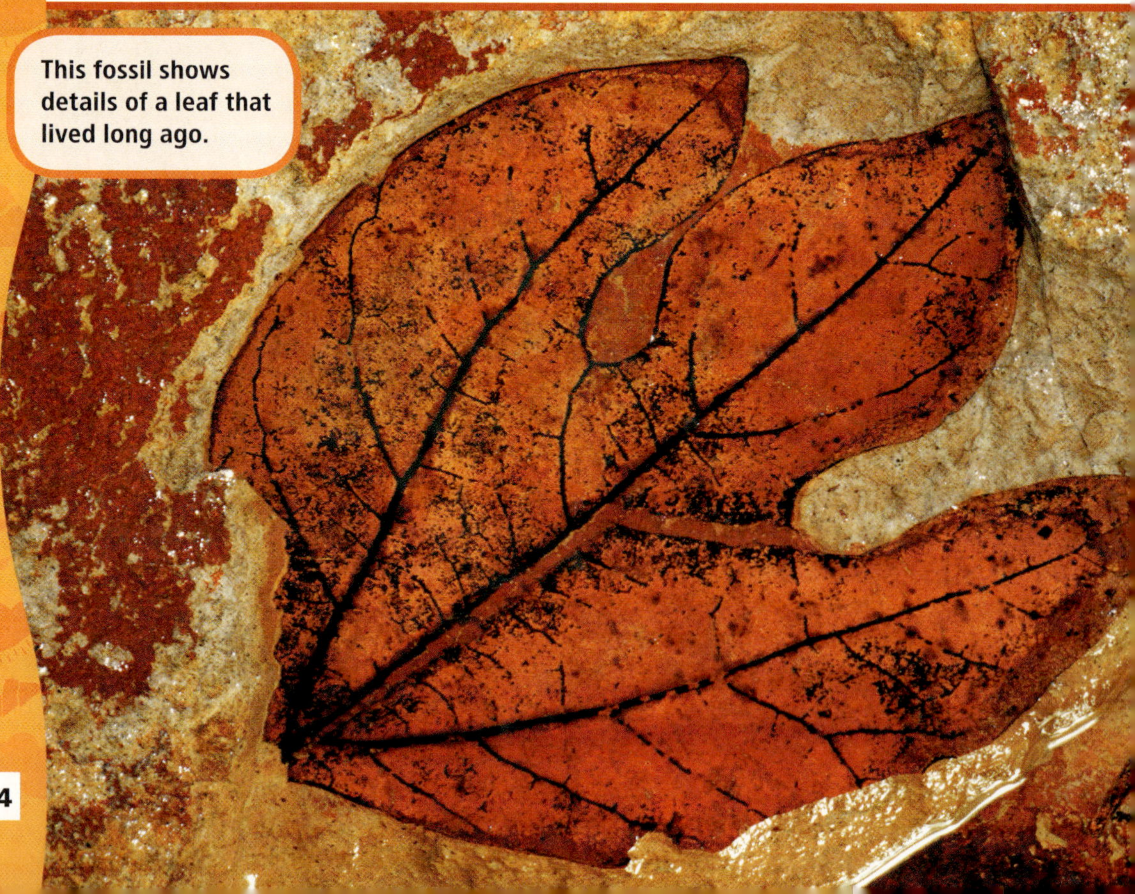

This fossil shows details of a leaf that lived long ago.

Studying fossils allows paleontologists to compare ancient and modern plants. Most ancient ferns didn't have seeds, and neither do modern ferns.

Summary

Fossils are the remains of past life. Studying fossils helps show what Earth's living things were like in the past. Fossils have helped scientists infer how animals, plants, and the climate and surface of Earth have changed over time.

Fast Fact

Fossils of ancient ferns have led scientists to conclude that these plants grew up to 18 meters (59 feet) tall. Some types of modern ferns also grow tall—up to 24 meters (80 feet).

Glossary

cast (KAST) A fossil formed when dissolved minerals fill a mold and harden (5)

fossil (FAHS•uhl) The remains or traces of past life, found in sedimentary rock (2, 3, 4, 5, 6, 7, 10, 11, 12, 13, 14, 15)

index fossil (IN•deks FAHS•uhl) A fossil of an organism that lived in many places around the world for a short period of time; it can help scientists find the age of a rock layer (10, 11)

mold (MOHLD) The hollow space that is left when sediment hardens around the remains of an organism and the remains then dissolve (5)